ARTIFICIAL INTELLIGENCE: FRIEND OR FOE?

Artificial Intelligence in Entertainment

Will AI Help Us or Hurt Us?

Nick Hunter

CHERITON
CHILDREN'S BOOKS

Published in 2025 by **Cheriton Children's Books**
1 Bank Drive West, Shrewsbury, Shropshire, SY3 9DJ, UK

First Edition

Author: Nick Hunter
Designers: Paul Myerscough and Jessica Moon
Editor: Sarah Eason
Proofreader: Kate Hobson

Picture credits: Cover: Shutterstock/Prostock Studio (foreground), Shutterstock/Dragon Images (background); Inside: Throughout: Shutterstock/Nadya Art, p1: Shutterstock/Gorodenkoff, p5b: Shutterstock/PopTika, p5t: Shutterstock/LightField Studios, p6: Shutterstock/Stokkete, p7: Shutterstock/4PM Production, p8: Shutterstock/Fizkes, p9: Shutterstock/Alessandro Biascioli, p10: Shutterstock/Gorodenkoff, p11: Shutterstock/John Doukas, p12: Shutterstock/VGstockstudio, p13: Shutterstock/Andrea Raffin, p14: Shutterstock/Zoran Zeremski, p15: Shutterstock/GuruXOX, p18: Shutterstock/Gorodenkoff, p19: Shutterstock/Pressmaster, p20: Shutterstock/PeopleImages.com/Yuri A, p21: Shutterstock/Gorodenkoff, p24: Shutterstock/Mike Dotta, p25: Shutterstock/Aslysun, p26: Shutterstock/G-Stock Studio, p27: Shutterstock/Halfpoint, p28: Shutterstock/KDdesign Photo Video, p29: Shutterstock/Monkey Business Images, pp32-33: Shutterstock/Rawpixel.com, p34: Shutterstock/Antonio Guillem, p35: Shutterstock/MDV Edwards, p36: Shutterstock/Nattakorn Maneerat, p37: Shutterstock/Gorodenkoff, p41: Shutterstock/Fizkes, p44: Shutterstock/Wachiwit, p47: Shutterstock/TippaPatt.

Printed in China

Contents

What Is Artificial Intelligence?

Have you ever watched a movie that has been created using **artificial** intelligence (AI)? You may not know it but there is a growing chance that you have. We're all familiar with the use of **computer-generated images (CGI)** for movie special effects, but AI is now offering moviemakers the chance to create effects that are much more **realistic** and are designed so audiences barely know that the movie has been changed. AI can make famous actors look younger or older to fit the part they are playing or to make last-minute changes to scenes that have already been filmed.

DEFINING AI

Before we look at the many changes AI could bring to our homes, what do we mean by AI? Artificial intelligence describes the way that **software** or machines can be designed and programmed to do things that can normally only be carried out by intelligent beings, including humans. These could include recognizing human speech, making decisions based on data, and carrying out complex tasks based on these decisions.

RECREATING EXPERIENCES

The world of entertainment includes movies, TV, music, and games. To some extent, in all these media, actors and other creators are trying to reproduce human behavior for our entertainment. Writers and movie directors try to create human stories that we can believe are real. Actors and musicians capture human experiences and emotions. Creating these moments requires great skill and years of practice.

Developments in AI have given movie studios, music producers, and others a new way of achieving the same things using powerful computers that can **replicate** these human experiences. AI is already being used to create new music and is being used in some parts of the movie industry—and this is just the beginning of the revolution in AI. AI is also changing the way we entertain ourselves as social media becomes an ever-bigger part of many people's lives.

AI tools are important parts of the team developing new movies and TV shows.

THE BENEFITS AND THE RISKS

Since the beginning of the twenty-first century, there have been huge steps forward in the development of AI. Many people believe that the changes brought by this technology are just beginning and will affect many areas of our lives. AI will bring major changes to the way entertainment is created and the way we access it. This will bring benefits such as more choice of entertainment that is closely linked to our personal preferences. However, it could also bring risks and problems for many, including the creative people who create the entertainment we love. For these reasons, not everyone is convinced that AI will be solely **beneficial** and should be handled carefully.

This book will look at how AI could change entertainment and **debate** whether this new technology is our friend or **foe**. Look for **IS AI A FRIEND OR FOE?** throughout the book. Read the arguments for and against AI, then answer questions that invite you to draw your own conclusions about whether this transformative technology will help us or hurt us.

CHAPTER 1

The History of AI in Entertainment

Most forms of **mass entertainment** have been invented since the early years of the twentieth century. The first moving pictures appeared in 1895 and the first golden age of cinema came in the 1920s and 1930s. TV followed from the 1940s onward. Other parts of the entertainment industry such as recorded music started to become popular around the same time. Computer-based entertainment such as video games became part of the entertainment scene in the 1980s.

IMAGINING THE AI WORLD

In the past, there have been many attempts by people in entertainment to imagine what a world dominated by AI might look like. TV shows and movies feature humanlike **robots** that can think and communicate just like us, but also have superhuman capabilities. While robots that move and look like humans are not yet a common feature of AI, scientists and developers have been working to create computers that appear to think like humans, but can do some things far more quickly and accurately than humans can.

AI IN THE REAL WORLD

Scientists first began talking about AI in the 1950s. Research began to design a computer that could think like a human, with some people **predicting** that this could happen by the 1970s. It later became clear that computers would need to be able to process information much faster than was possible at that time.

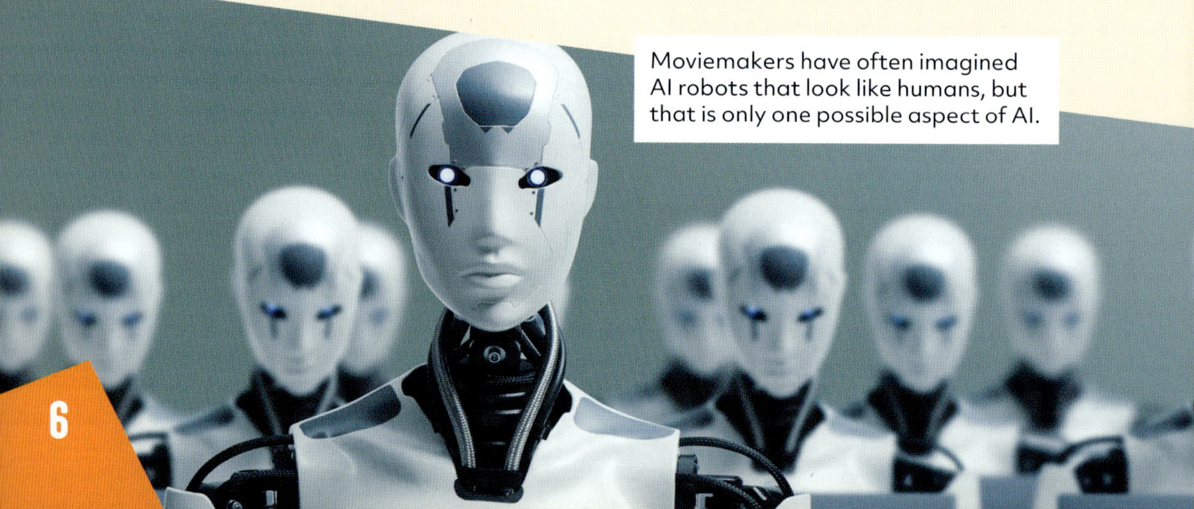

Moviemakers have often imagined AI robots that look like humans, but that is only one possible aspect of AI.

AI systems are trained using the vast amounts of data we send and access via the Internet.

Development of true AI has been made possible by advances in the speed of computers, using **microprocessors** that were first designed for online gaming.

INTERNET ENTERTAINMENT

Since the 1990s, the Internet has changed the way we access our entertainment. We might still go out to a movie theater to catch the latest blockbuster, but we increasingly watch movies from the comfort of our homes via **streaming** services. These services give us a huge choice of what we can watch and allow us to consume entertainment whenever we want.

Big Data

The Internet was essential for making progress with AI tools. AI appears to make decisions and plan like a human, such as in understanding the human voice and responding to spoken language. In order to achieve this, AI systems have to be trained on huge amounts of online data, such as the text of millions of books or articles so that it can spot patterns and learn from this material.

Steaming services give us access to a vast range of music, and AI helps guide what we choose to listen to.

MEDIA CHOICES

Streaming services have also changed the way we listen to music. Rather than buying music in a physical form on record or CD, we have access to a huge range of music choices via smartphones and other devices. The Internet also gives us access to many games, magazines, books, and other media.

PREDICTING WHAT YOU WANT

AI has been part of the media revolution for some time. Your favorite movie and music streaming services will recommend media that they think you will enjoy based on what you have watched in the past or other data they have about you. They may not always get it right, but these recommendations are getting more specific all the time. The more data that AI has about you, the more accurate it should be in guiding your choices. Music playlists are generated by AI to group the most popular songs by an artist, or songs that suit a particular mood.

WINNING WATSON

In 2011, the world of AI and entertainment came together. IBM developed an AI computer named Watson that was able to beat two champion quizzers on the TV game show *Jeopardy*. The amazing thing about this was not necessarily that Watson knew the answers to the questions but that the questions (or answers) were often phrased in natural language that was not easy for a computer to understand.

CHANGING PRODUCTION

Since 2000, AI has seen growing use in producing entertainment. This could include changing a singer's voice or the sound of an instrument in music production to using AI as a quick way to edit the scenes in a movie or TV show.

NEW POSSIBILITY

The growth of **generative AI** has opened up new possibilities for the use of AI in entertainment. The AI is trained on millions of documents or images. Tools such as OpenAI's ChatGPT can use this data to generate new media. This could include text to answer a specific question or to replicate the style of a particular author. It could also include images or music, all using the style of a particular songwriter or the voice of a successful singer.

CHANGE AND CONCERN

As we will see, recent advances in the world of generative AI could completely change the way some entertainment is produced. However, many people are very concerned that the advances will damage the rights and earnings of writers, actors, and musicians.

Digital technology has already revolutionized the way that music is made, and AI technology could have an even bigger impact.

CHAPTER 2

Movies and TV

Do you know how your favorite movie was made? It has been a long time since most movies were made by simply pointing a movie camera at a group of actors. Big-budget movies and TV shows often involve hundreds of people. Many of these people are working with technology to create **movie sets** and special effects. Today, actors often perform their roles against plain backgrounds with sets, special effects, and other characters created on computer and added later.

ACTORS GROW OLD AND YOUNG

AI is already adding new elements to this process that change the way movies are made. Actors can now be altered by AI. When a movie studio is looking for actors for the latest blockbuster, the age of the actor may not be perfect for the role. Will people believe that a famous actor who has been on our screens for many decades could play a character in his 20s or 30s? Some movies need the actors to appear at different ages in the same movie, and AI can help with that.

The way a movie is shot may not have changed too much, but today AI is often used to alter how the actors appear on the screen.

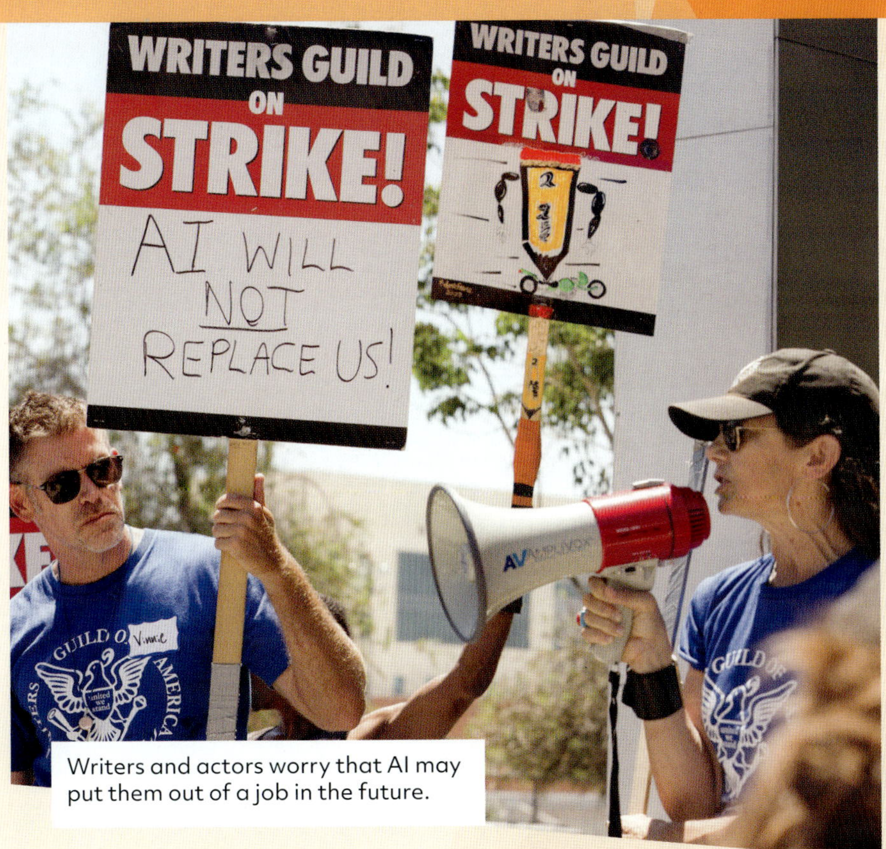

Writers and actors worry that AI may put them out of a job in the future.

AI APPEARANCES

AI has been used to make actors appear younger than their real ages. A younger version of the actor's face can be created with AI and then added to the digital files of the actor's **original** performance. Using AI to make changes like this is cheaper and more realistic than using previous CGI techniques.

VISUAL EFFECTS

Producers of visual effects are also using AI to make **special effects** more realistic and to reduce costs. Visual effects producers expect that AI will not take over from CGI and other techniques completely, but it will be one more tool they can use to make visual effects better. In time, all visual effects could be produced by using AI technology.

SCREENWRITING

In 2023, writers in Hollywood went on **strike**. This was a disaster for the movie industry. Without writers, there are no new ideas for the hundreds of other people involved in producing a movie to work on. Without writers, actors would have nothing to say. One of the key reasons for the writers' strike was the growing threat of AI technology to their work. Many people in the movie industry joined the strike.

Screenwriters start with a blank screen or sheet of paper and create blockbuster movies and characters.

CREATING SCRIPTS

New AI tools can analyze writers' **scripts** and ideas and suggest changes, but writers are worried about what could happen next. Imagine you are a writer of a TV drama or comedy. In every episode, the characters face a different problem, such as a crime to solve. By the end of the episode, the case is solved and the story concluded.

THE NEXT SUPERHERO

Currently, each episode usually needs a writer or a team of writers working on it. Advances in generative AI mean that new stories and even whole scripts could be created by AI, which would learn from previous episodes and information about the characters and other elements of the show. If that's possible, maybe AI could create the script for the next part of a superhero movie series.

NOT NEW IDEAS?

Writers are understandably worried that AI could put many of them out of a job. Scripts created by AI would be cheaper and created much more quickly. Maybe the rest of us should be concerned too. Writing a screenplay is not the same as writing a business letter or a report. It needs **creativity** and that is something that AI finds much more difficult. We should be concerned that TV or movies written by AI would not have the creativity that keeps us watching. They might all start to look very similar!

ARTIFICIAL ACTORS

Actors too are worried about what AI could mean for their work, and have also organized strikes in protest. The biggest Hollywood stars worry that AI could be used to replicate their image and voice and create fake versions of them, which they would be unable to control. For example, actor Stephen Fry discovered that AI had recreated his voice using recordings he had made for **audiobooks** of the Harry Potter stories. The **cloned** voice was then used to **narrate** a **documentary** film without his **permission**.

CLONING CROWDS

Many movies and TV shows also involve a lot of additional actors who are not instantly recognizable. Some of these supporting actors may simply be faces in a crowd or battle scene. They too are worried that their images could be cloned by AI and used in movies without their permission and also without them being paid. This could be an attractive idea for movie companies because it would help them save time and money. They would have to hope that viewers will not be able to spot the difference!

The careers of movie stars such as Leonardo DiCaprio depend on their faces and their voices. Movie stars are therefore rightly worried that these could be copied using AI.

Researchers are currently using AI to test what will make us laugh or cry at the movies.

BLOCKBUSTERS

Hollywood studios are not just using AI to make movies, they are also using it to help them determine which movies will be successful. One of the big benefits of AI is that it can analyze a lot of data quickly. This means it can use data about different moviegoers, their likes, dislikes, and what they have watched before, to predict whether a movie will be a success. This knowledge can be used to advertise new movies and guide which movies are made in future.

FITTING A FORMULA

If AI can tell so much about the movies or TV shows that people want to see, this could make the movies produced much more similar to each other as they would all fit a successful formula. If we know what type of characters and storylines are likely to be successful, this could mean that many creative ideas are dismissed as unpopular. That would mean far less variety in the movies and TV shows that are made. Is that a good thing?

JOB WORRIES

Writers and actors are not the only people concerned about the effect that AI could have on Hollywood. In a 2023 survey, 42 percent of professionals working in the entertainment industry said they were worried about the effect of generative AI on their work. These people are concerned that AI could take over many processes in movie and TV production. Movie and TV professionals recognize that their industry could be one of the first to be changed by generative AI.

COPYRIGHT CONCERNS

The use of generative AI in entertainment raises lots of legal and **ethical** questions. AI tools can only be trained using data that already exists. This data could be made up of the words written by authors and screenwriters, millions of images of actors, or images from other movies. AI then uses this data to create modified versions or entirely new images. However, the people who created the original words and images are angry that their work is being used to train AI, and that very technology could then put the creators out of a job.

GETTING CREATIVE

Many people also believe that use of AI may be cheaper and more efficient for making movies and TV, but the most important skill in making entertainment is creativity. Writers can have original ideas that we want to explore. Moviemakers and actors can bring them to the big or small screen. Many people believe that AI will never be able to recreate the excitement we get from great creative entertainment.

There are many technical roles in Hollywood that could be affected by developments in AI.

The Debate:

AI Can Replace Movie Actors

Movie history is full of brilliant star actors who cast a spell over their audience, but could the age of the movie star be coming to an end with the growth of AI? Perhaps AI will replace star actors and ordinary actors too. There are arguments both for and against this scenario. Let's take a look at them.

AGREE

AI will save time and money: Movie stars are expensive and making movies takes a lot of time. If producers can use AI to save money and people can't tell the difference between real actors and AI, they will make that choice.

Some actors are easier to replace: We may know the names of the stars in our favorite movies, but many actors may appear only in the background for a few seconds and are often barely noticed. These roles would be easier to replicate with AI images.

Some movies don't need real actors: Many blockbuster movies are all about CGI special effects and spectacle. Do these movies really need the skills of a human actor to make them popular? Many people argue that the draw of such movies is their impressive action scenes, not the movie stars themselves.

DISAGREE

Actors bring creativity and emotion: Acting is about more than just speaking the lines in a script. Actors can bring emotion, humor, and all the little details that bring a character to life. AI will never be able to replace these skills. Only people can add the detail that counts.

Star power makes money: There's a reason why some stars' names are so visible on the movie poster. They are a big reason why we might choose to watch one movie instead of another. That's why they earn big money too. AI does not have this star power.

Legal and ethical issues: AI tools have to be trained on real data, such as the look and sound of real actors. Copying a real actor's face or voice could create legal problems. The idea of AI replacing actors also raises ethical questions. Is it really fair to replace humans?

Conclusion

The growth of AI will create challenges for many actors who work in movies and TV. However, AI is unlikely to replace major stars, although they may have to deal with their images and voices being replicated. It could have a greater impact on jobs for less well-known actors. It is another question entirely whether the use of AI to replace humans will improve the quality of entertainment, and some people think it won't.

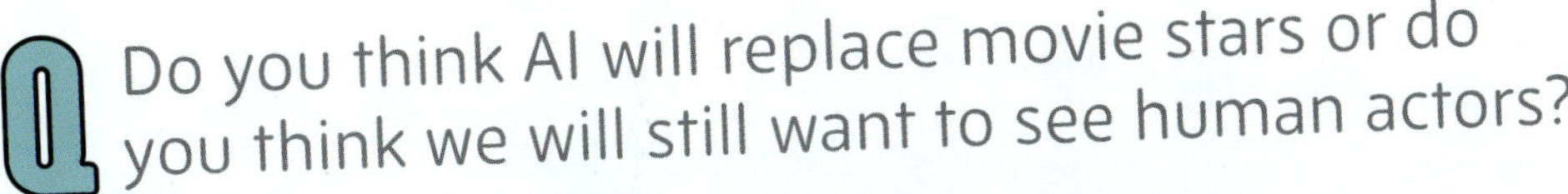

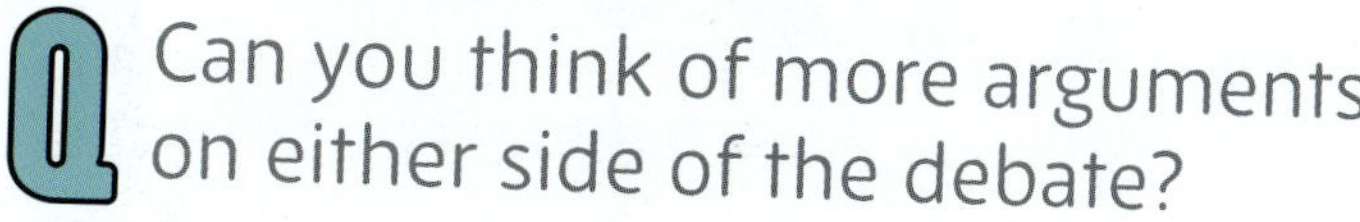

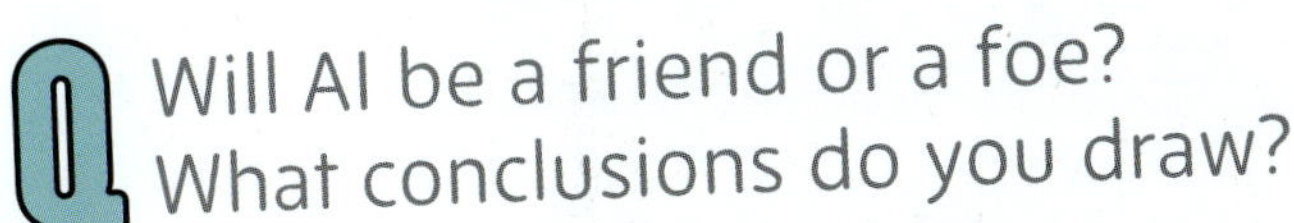

CHAPTER 3

Making Music

Music can have a magical effect when we hear a song that we love. However, when we stream music on to our smartphone or computer, it's just a stream of data such as text or a picture. Each piece of music has a digital fingerprint—that's why a music recognition app can tell us what song is playing from just a short extract. If music is data, then AI can be trained to recognize and replicate music. This can be useful in many ways, but it can also create certain problems for musicians.

MUSIC PRODUCTION

AI can help musicians build songs from scratch. Just as generative AI can create a document or image based on the data that has been used to train it, AI music tools can create new tunes for musicians. Apps exist that enable inexperienced musicians to select a beat or song style and AI will create a tune in that style. The musician can then mix up the different elements or record their own vocals or instrument to accompany this tune. It's an easy way to make music, but the AI is doing most of the composing rather than the human musician.

Skilled producers can combine the work of human musicians with sounds generated by AI.

AI could replace human musicians in creating music for soundtracks and advertising, which musicians often depend on to earn their living.

BREAKING AND REMAKING

If AI can build new music, it can also separate existing music into its different parts, by recognizing which notes are produced by the different instruments. That means if someone wants a version of a song to include on a movie soundtrack, but without vocals, this can be created without any involvement from the original musicians.

OLD GROUP, NEW SONG

The Beatles are one of the most popular musical acts of all time. The group broke up in 1970 and two members later died. In 2023, the surviving members announced the release of a new song featuring all four musicians. Existing music had been separated using AI and then new material was added to create a new song from a group that could no longer record together.

CREATING NEW MUSIC

The ease with which AI can create music leads to the real possibility that it could create new music that builds on the work of other musicians. Providing music for movie soundtracks can make a lot of money for musicians. This money could disappear if movie companies can easily create a tune in a style that sounds exactly the same as an existing song but is slightly different from the original. That possibility is a concern for artists.

NEW VOICES

Many tools are being developed that can create new music based on the work of other musicians. This is all being done by AI with no human musicians being involved. As well as movies, this music can be used as a soundtrack for video games, social media video clips, and much more.

EXISTING VOICES

AI can also replicate the voices of existing performers. This could lead to versions of old songs with new vocals that are an exact copy of the biggest music stars from the present or the past. Songwriters can record a version of their song with the voice of a major artist to show what it could sound like if that artist recorded the song. A company may not be able to persuade Taylor Swift to sing on their latest advertisement, but they could create a music track that is close enough in sound for people to believe they are hearing a star singer. That big jump in music creation would have a significant impact on major recording stars.

Music can trigger strong emotions of joy or sadness. Will the music created by AI have the same effect on people?

The exciting experience of seeing and hearing a band playing live is something that AI cannot replace.

MUSICAL DIFFERENCES

While musicians may enjoy the benefits of using AI to write or perform their next hit, artists and music companies are worried about the ease of using AI to replicate music. Musicians earn money whenever their music is played in public, whether it's part of a TV show or playing in a shopping mall. AI-created music could have a big impact on these means of earning money for musicians everywhere.

MISSING CREATIVITY

Critics of AI-created music point out that it is just a very good copy of existing music—all the same notes but not necessarily in the same order. A great vocal or musical performance is full of emotion that connects directly with the listener. AI will never be able to reproduce the creativity and emotion of the human musician. As AI develops, producers will try and find new methods to match the emotion of a human performance. As it is, if music produced by AI becomes the main way that music is created, we will all miss out on the thrill of listening to great human musicians.

COPYRIGHT CRISIS

Copyright is the law that stops others from copying or stealing an artist's work. The music industry is likely to see many disputes over copyright as musicians try to stop AI from copying their voices and styles of music. If they are not successful, we run the risk of losing many of the great performers who make the music that is such an important part of our lives.

The Debate:

AI Will Improve Music

Musicians can see that AI could bring new opportunities to create music, but are worried about the idea of AI making music without the need for musicians at all. Perhaps AI will create our music, and even improve music-making. There are arguments both for and against this scenario. Let's take a look.

AGREE

Building on the past: AI systems can learn from all the music that's been made in the past to come up with new compositions, just as human musicians learn from the music of people that influence them.

Supporting musicians: AI can support musicians in creating new music and save them time. For example, AI could create a backing track so the musicians can focus on other instruments or melody. A musician would normally have to pay supporting musicians to do that type of work for them, but AI can do it quickly and for free.

No need to be original: AI will mostly be used to create music that doesn't need to be very original, such as backing music for video games and advertising. Human musicians will be able to focus on something more interesting, such as creating new music or coming up with a fresh, new approach to the way they play their music.

DISAGREE

AI can only learn from existing music: AI can make very complex music and mimic existing styles and voices, but it is much less successful at creating something that has not been done before.

Music needs human emotion: Human players and singers make music that inspires and moves us. The best music is unique to the people who write or play it and AI will never be able to completely replicate this. Only humans can create exciting original material.

Real musicians will not be able to compete with AI: If more music is created by AI, this will make it more difficult for real musicians to be successful. Being a musician requires talent and a lot of practice. If there are fewer musicians, that means less original music for us all.

Conclusion

It's difficult to believe that most people would choose to listen to music made by AI rather than music made by human musicians. But much of the music we hear in advertising, on a game soundtrack, or in an elevator is not chosen by us. AI **algorithms** will need to develop before they can produce original music, but this may come in the future, and at that point we may find AI creates much of our music.

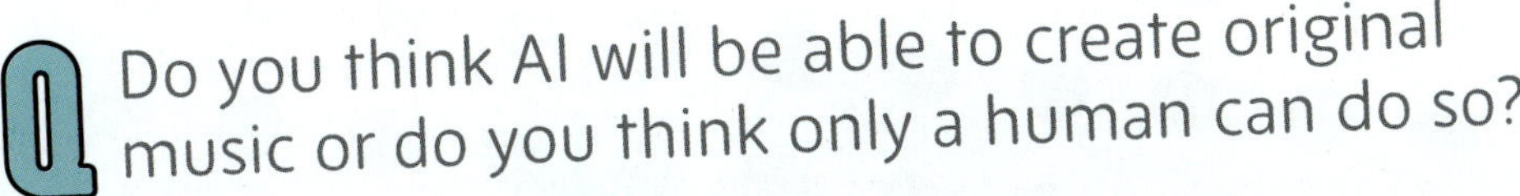

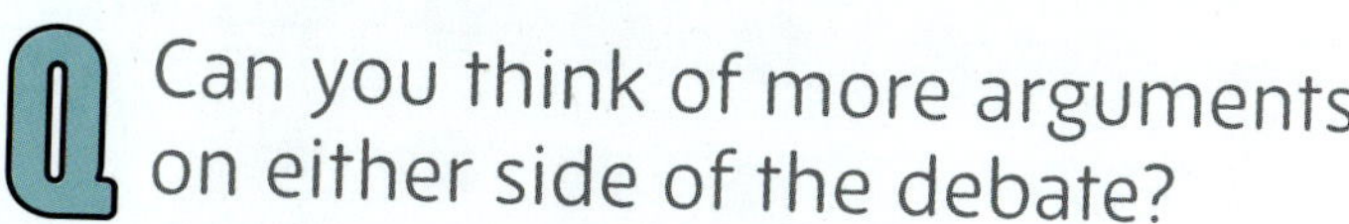

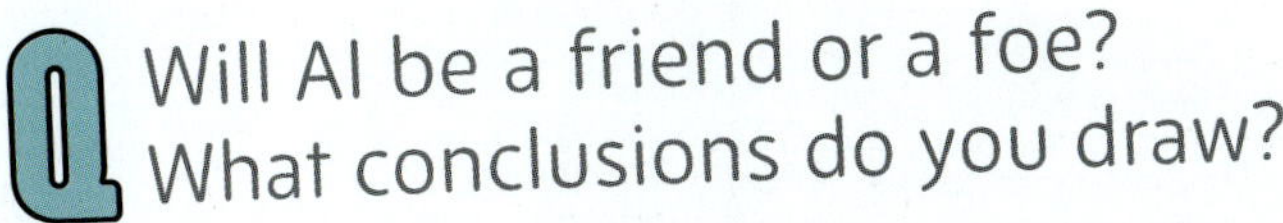

CHAPTER 4

Gaming and a New Reality

Online gaming has become one of the most important areas of the entertainment industry. Experts have estimated that one out of every three people on Earth played a video game in 2021, and games are still becoming more popular. When the Covid-19 **pandemic** forced many people to stay at home for long periods, online gaming saw big increases in player numbers as people of all ages played games for enjoyment and to help them connect with their friends online.

NEW HARDWARE

Gaming is a natural area in which AI can have a strong influence. The microprocessors that are used in gaming have been important for developments in AI because they can process huge amounts of data at high speed. AI is already starting to influence the **hardware** we use for gaming. In fact, Google recently demonstrated a mouse named GameFace. It is controlled using only the player's head movement and **facial gestures**. It was inspired by the experience of a gamer who could not use his hands.

A MAGIC MOUSE

The mouse uses AI machine learning to learn and recognize different facial expressions and movements and link them to different mouse commands.

This application of AI can be used more widely than just for gaming. It is designed for the speed of reactions needed for many games.

REALISTIC GAMES

AI will undoubtedly have a big impact on games themselves to make them appear more real and **immersive** for the gamers. Many games already use AI and developers are always looking for new ways to make games more

Games are often used as a way to test and develop new hardware and software.

Pro gamers will be challenged by the use of AI to develop characters and environments in new games.

realistic. Use of AI will enable developers to fill their games with different characters, more like the real world. AI will give enemies or nonplayable characters (NPCs) you meet in games more complex personalities. Many of these characters will feel as if they are being controlled by another player, with some being more cunning or others looking for revenge. The game and the characters could change depending on what you do. They can also communicate with you in natural language, rather than just using a few different commands or phrases.

Making Games More Accessible

GameFace is one example of how AI can make games and technology more accessible to people with disabilities or who find it difficult to use technology for any reason. AI is able to complete the link between a gamer's head movements or voice and the commands needed to play the game. Developers are also working on tools that will describe worlds in games such as Minecraft to make them more accessible for visually-impaired players.

The jobs of game developers will change as AI becomes a bigger part of the process of developing and creating online games.

CHANGING GAME DEVELOPMENT

AI could certainly speed up the complex and expensive process of game development. Making games requires a lot of work by large teams of developers, creating the characters, backgrounds, and features that are essential for the gaming experience. Some games need many hours of music and voice recording. All this costs a lot of money and explains the high price of many games.

AI WORLDS

If AI can develop the most time-consuming aspects of games, developers' roles could be more about improving and refining the worlds created by AI. Settings for games will also be more realistic with AI adapting and developing settings as the game continues. AI could learn the structures of real settings, such as deserts or city streets, and use them to create its own realistic environment. AI can also create storylines based on the characters in the game, so the possibilities for adventures become endless.

CREATIVE WORRY

Just like movies and music, the best games depend on human creativity. As AI is able to take more control of games, this could lead to less human involvement. This could be bad news for games developers, who may find their jobs being taken by AI. It could also have a negative impact on gamers' experiences. Without human creativity, games could lose some of the artistry that makes them so exciting for many people. The growth of huge games controlled by AI may mean that there are fewer new games.

TOO DIFFICULT

Another concern is that games controlled by AI could just be too hard. AI characters may be able to learn what the gamer is going to do before he or she can do it. We want games to be as realistic as possible but it's no fun if we're always losing.

A NEW WORLD OF ADVENTURE

Use of AI in virtual reality (VR) could create a whole new world of games and entertainment. VR usually involves wearing a headset that can take us into a new artificial world. Advances in AI could make such a virtual world almost unlimited because AI can use data and images of the real world to create another world where we can meet **avatars** of other people and enjoy new entertainment experiences in what has been called the "**metaverse**." This development could also be part of the future of social media.

Experts predict a big expansion of the experiences that will be available through VR.

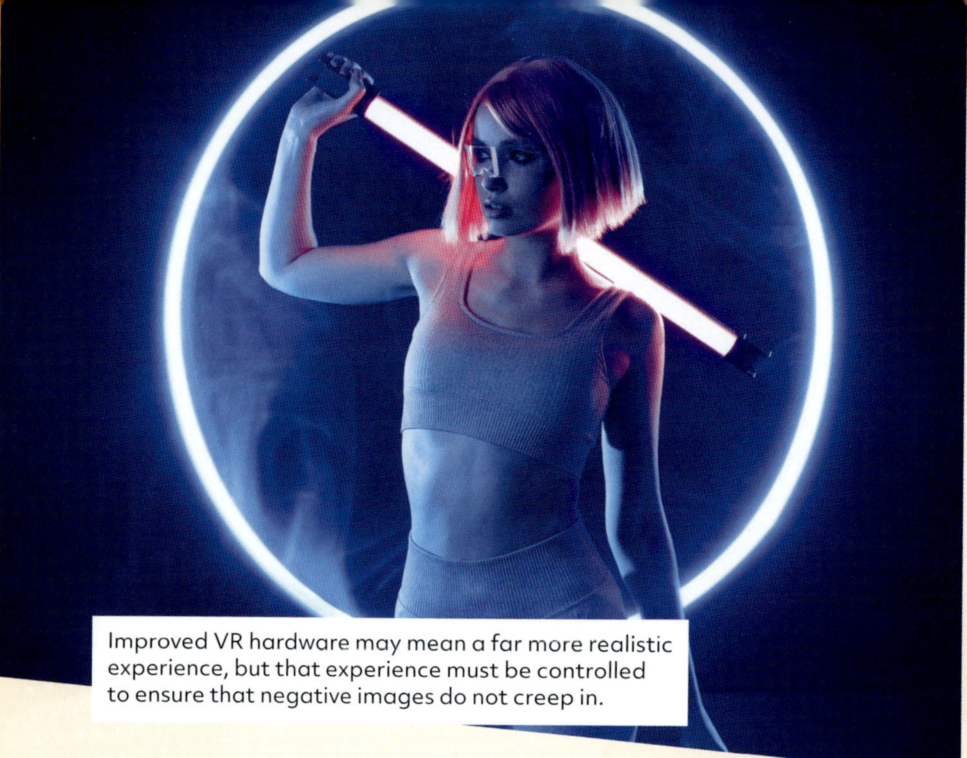

Improved VR hardware may mean a far more realistic experience, but that experience must be controlled to ensure that negative images do not creep in.

REAL-LIFE ISSUES

If AI is going to make our games more immersive and more exciting, what could be the problem with that? One possible concern is the fact that, as games become more realistic, they may also include some difficult real-life issues.

LOOKING FOR PATTERNS

In order to create realistic worlds and characters, AI is trained by looking for patterns in huge amounts of data. This could include millions of images, videos, or words spoken by real people. The risk is that games will replicate some of the worst aspects of the real world. This could include repeating **stereotypes** of different groups in games, showing some groups of people in negative roles, or games excluding strong female characters. While game developers may have some control over the content that goes into games, it is much more difficult to ensure AI does not repeat **bias** or **discrimination**.

PROTECTING PRIVACY

Privacy is already an issue for online gamers. We always need to be careful not to give personal information to people we meet online as we cannot be sure who they are. AI could enable people to create more realistic avatars online. These realistic avatars could hide the identities of people who would try to do us harm, persuading us to trust them more than we should. That could be extremely dangerous.

DANGERS ONLINE

There is also the danger that as we spend more time in online worlds, we will reveal more about ourselves. This could give other people the information they need to commit crimes. Even if we are not being misled by cybercriminals, we should always be aware that our personal data may be being used by gaming companies and their AI tools to find out more information about us.

GOING TOO FAR?

A final concern is that these new AI-controlled games and virtual worlds could prove too attractive. There is evidence that too much gaming can have a negative effect on people's lives and that game addiction is a problem for some people. There is a risk that the benefits that AI can bring to gaming will make these problems worse. People may find the alternate reality of their favorite game more exciting than real life.

Studies have shown that too much gaming can affect a person's sleep and mental health.

See pages 30-31 for more on the debate about gaming and mental health.

The Debate:

AI Will Make Gaming Better

AI has the potential to make gaming far more exciting for many of us. Perhaps it will improve the gaming world and enhance the experience for everyone. There are arguments both for and against this scenario. Let's take a look at them.

AGREE

More realistic characters: AI will enable characters within games to be much more rounded and interesting than ever before, so it will feel as if you are competing with a lot of other human players in every game. That will ensure a much more realistic and exciting gaming experience for all the players in the game.

Developing better games: Developers will be able to use AI tools to develop games more quickly, and in the future AI will create more levels and features so gaming worlds will be more exciting.

Accessibility: Developments such as controlling a player by head movements or voice control will help make games more accessible for players with disabilities or those who find gaming difficult.

DISAGREE

Less human creativity: More use of AI could mean that there are fewer human developers involved in game design. This could lead to less room for creativity in developing games so future games are less exciting and feel more like they were created by a computer.

Too much reality: AI could make games too much like the real world, which would bring real-world problems such as bias and discrimination into games. AI could also create more opportunities for criminals, who use the games to try and mislead the gamers.

Risk of addiction: Games of the future could become so realistic and immersive that there is an increased risk of health issues and even addiction for people who spend too much time gaming.

Conclusion

The use of AI in developing games could make for more exciting experiences and make games easier for many people to access. However, this also brings many potential problems that will need to be solved to improve the gaming experience for everyone.

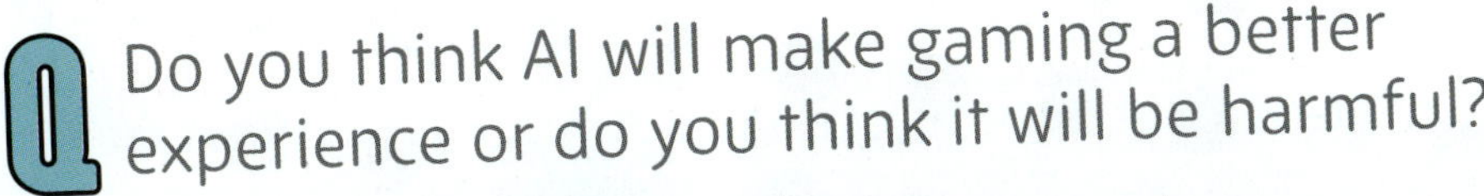

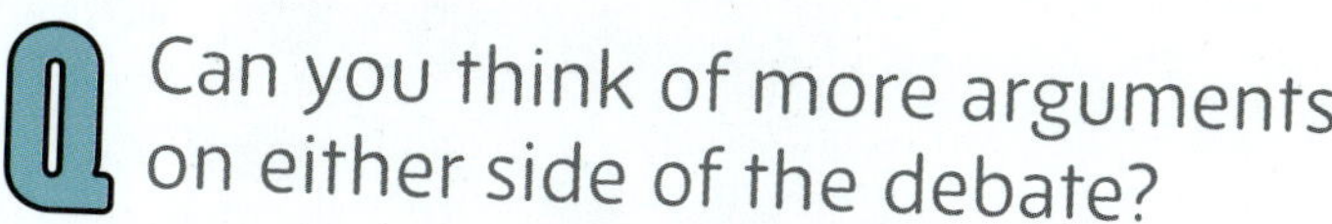

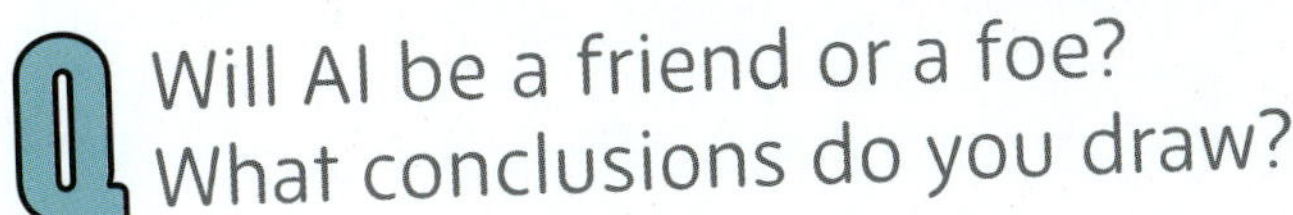

CHAPTER 5

Social Media

Social media has been part of the world of entertainment for far less time than other media such as movies and music. The first social media platforms developed in the early 2000s. The most popular platforms now have hundreds of millions of users. They combine parts of all other media as users can create and share videos, music, writing, and more to connect with and entertain people.

THE AI MACHINE

AI is already a major part of social media and it is no surprise that big social media businesses such as Meta, which owns Facebook and Instagram, and YouTube owner Google are some of the biggest players in the AI revolution. From the user point of view, social media is about connecting and sharing information with friends. Many people also get much of their news from social media. For the businesses that provide us with free platforms to do this, social media is about collecting data about millions of people and using this to target advertising and other services.

COLLECTING DATA

Everything we do on social media is collected and measured. What do our profiles and posts say about us? What things do we like? What videos or photos do we choose to look at? How long do we spend looking at each post?

DATA DECIDES

Mapping these interactions is made possible by AI systems that can sort data into categories and predict what this says about each user. They can then use this data to decide what we see on our social media feeds. The idea is to keep us scrolling and clicking for as long as possible. The longer we spend on social media, the more chance there is for advertisers to sell us things and the more chance there is to collect even more data about us.

AI COMPANIONS

The next step from a personal feed is for AI to provide us with a personal social media assistant. There are already services with which users can create a personal AI friend who they chat to online. Even though they know that their friend is artificial, thousands of people find that conversation with an advanced AI **chatbot** is not too different from chatting to a real human. These online companions will be armed with all the data about your life, likes, and dislikes and will be able to chat with you as if they are your best friends.

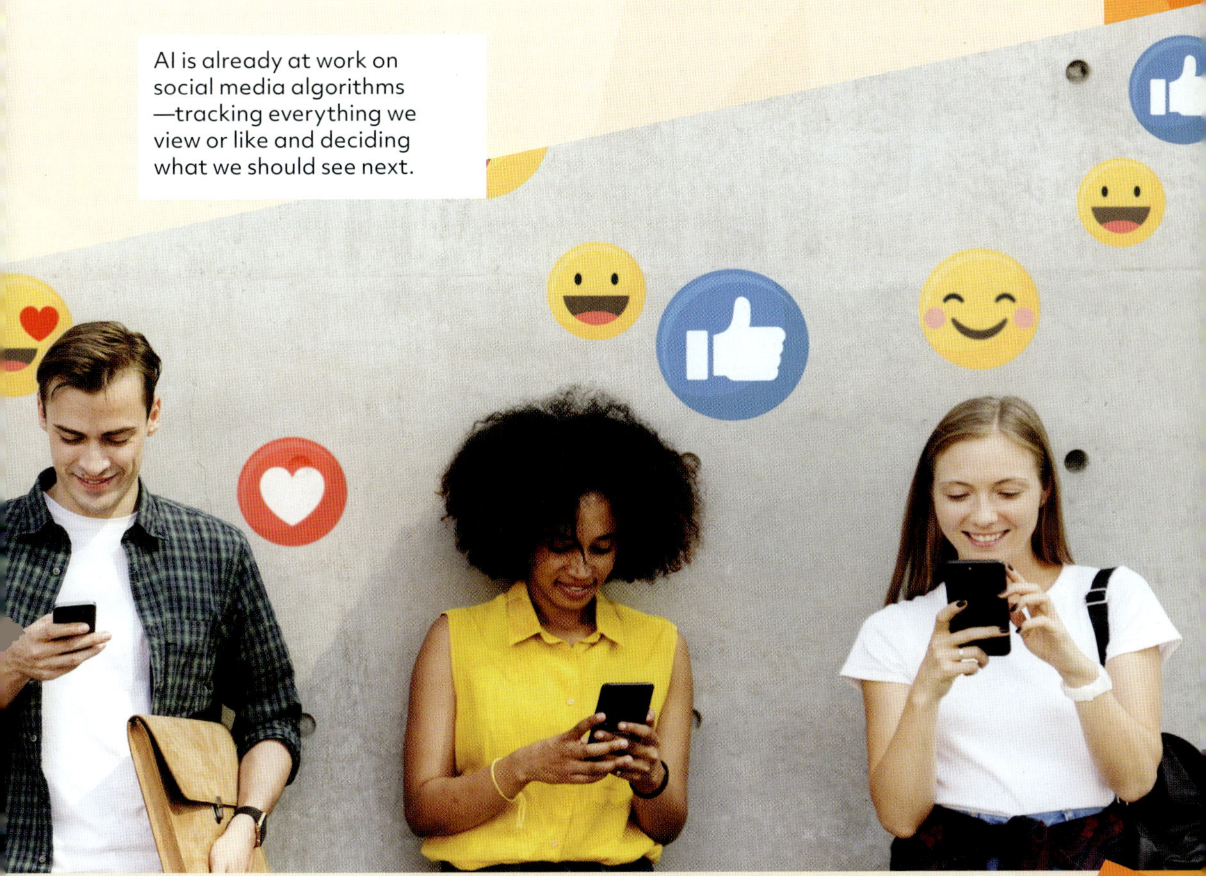

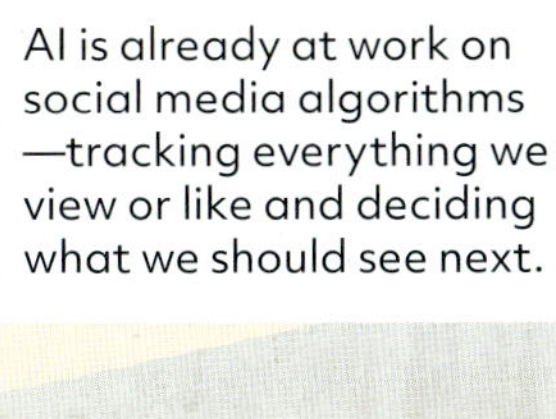

AI is already at work on social media algorithms—tracking everything we view or like and deciding what we should see next.

GENERATING CONTENT

AI has been deciding what content we see for a long time, but now generative AI can create media based on what we and people like us have looked at before. Social media has always enabled users to share their own photos and media. AI could now help us create high-quality text, images, videos, and music. This will enable all of us to create all kinds of posts just by typing a few words into an AI tool and seeing what it creates.

SMART FAKES

The downside of everyone being able to instantly create social media content is that social media feeds will be flooded with new posts and images. While a lot of it may be true and useful, some of it will be untrue or misleading. AI can create images that are known as **deepfakes**—meaning that it is very difficult to tell if they are real or fake. It is already very difficult to tell what's true on social media and the growth of AI-generated information could make this even harder.

OUT OF CONTROL

Even some AI insiders are worried that AI systems have started to do things that they have not been taught to do or were not part of their original planning. This is worrying because these systems are so complex that it is almost impossible for human developers to correct the problems. There is no single system of computer code that can make the fix as AI systems are continually making new connections. One result of this is that AI systems can "**hallucinate**," or generate content that is incorrect or untrue. This is worrying enough, but there is also the problem that humans can create misleading or dangerous content using AI.

You can expect to see more fake content on your social media feeds in the future.

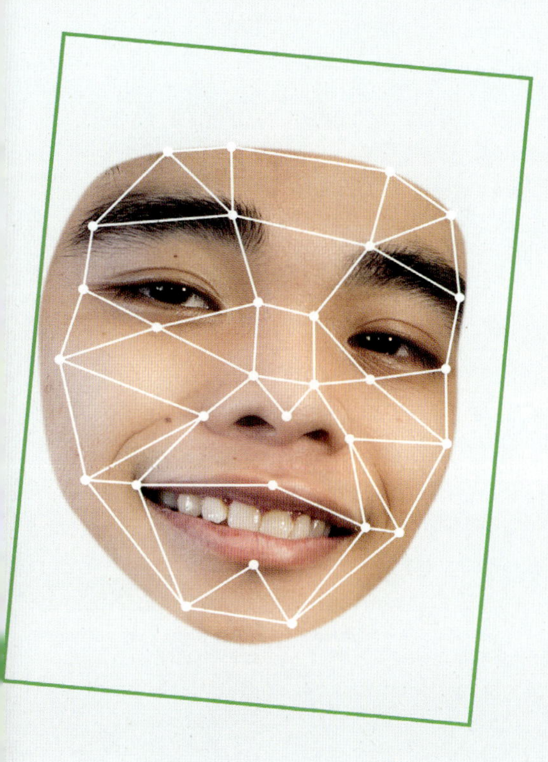

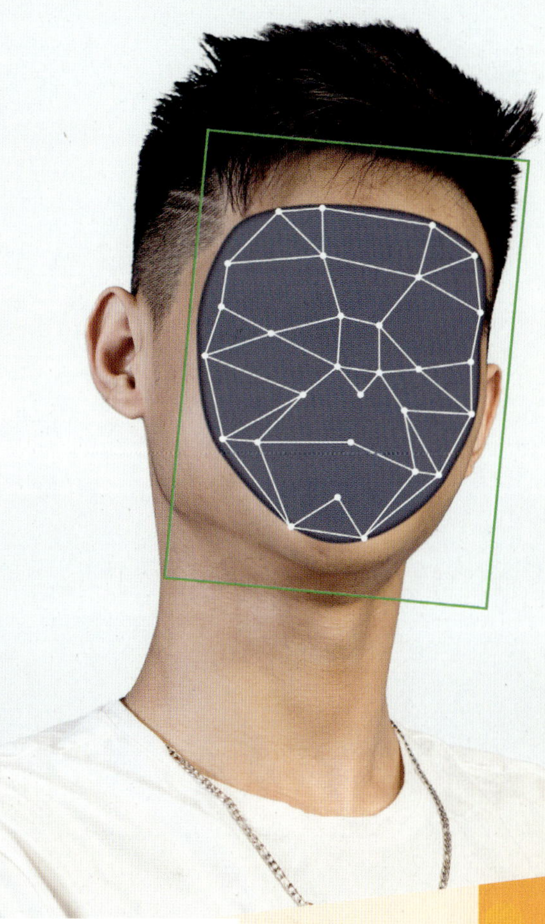

Voices and faces cloned by AI can be used to create fake social media content.

Creating Deepfake Images

The way AI creates new images is changing all the time. AI systems are trained on hundreds of millions of images, with text tags describing each image. The image is then broken down into a small collection of pixels so the AI can understand the different elements that make up the image. When someone asks for a new picture of, for example, a cat on the moon, the AI can use it's training to put together the different parts needed to create that exact image.

See pages 38–39 for more on the debate about AI and creativity in social media.

Social media companies want to keep you online for as much time as possible, and AI can help with this.

SOCIAL MEDIA PROBLEMS

In 2023, a school board in California filed a legal case against some of the world's biggest social media companies. They claimed that the companies were "using perhaps the most advanced artificial intelligence and machine learning technology available in the world today" to purposely design products that are "addictive and deliver harmful content to youth." The school board's case argued that social media was deliberately designed to affect young people's mental health and damage their education. Their concerns are felt too by other organizations and many parents.

KEEP ON SCROLLING

There is no secret that social media is designed to keep us watching and scrolling for as long as possible. AI will give social media providers powerful new ways to keep us hooked. TikTok is a social media site that is particularly popular with young people. TikTok's algorithms make use of AI to make every user's feed very personal. This is very successful in persuading users to spend time on TikTok. Other social media are taking the same approach. When much of the media is actually created by AI, this will make it even more attractive and possibly addictive to users too.

FAKE NEWS

Many people are worried about the way fake information and images can be created by AI. These could be used to influence people unfairly, to commit crimes, or to win an election. Campaigners believe that it is important for people to be informed about these aspects of AI. Social media enables untrue information to spread around the world very quickly, so it is important that there are international rules to stop this.

SOCIAL SPIES

Governments can also make use of AI in social media. Many social media providers insist on having no government interference in their services. This can cause disputes when governments try to tackle illegal content or other criminal activity on social media. However, in some countries, governments use social media data to spy on what their citizens are doing.

AI CONTROL

AI can study the social media activity of millions of people to spot any posts that are critical of the government. The global reach of social media can also be used to gather data on citizens of other countries. For this reason, Chinese-controlled TikTok is banned on US Government devices.

The data we create on social media is vital for corporations and governments.

The Debate:

AI Will Make Social Media More Creative

Many people who post writing, photos, and video on social media see themselves as creators. Some people are able to make money as YouTubers or influencers. Most people are happy just to show their creative side to friends. Perhaps AI will enable everyone to share more creative content. There are arguments both for and against this scenario. Let's take a look at them.

AGREE

Creative tools for all: In the past, people had to have the right technology to shoot high-quality photos and video. Smartphones have changed this already, but AI tools will give more people the tools to create new images, videos, and music.

Learning from the best: We all learn from looking at what has been done before. AI is trained using millions of images and other media, so it can create new materials based on the best examples from the past. AI is just improving on past material.

Social media becomes more personal: Use of AI in social media algorithms has helped people find content that matches their interests more closely. This should help newly created media to reach an audience that it otherwise might not be able to reach.

DISAGREE

AI is creating the content: If media is being created by AI, how much human creativity is involved? We might argue that AI created the idea, but a poet or an artist uses their skill to bring the idea to life. It is that human element that is creative.

Algorithms control what we see: Social media algorithms are not designed to show us the best creative content. They are programmed to keep us on social media as long as possible and provide data for advertisers and others who want to influence us. Many people feel there is no creativity in that.

Deepfakes and hallucinations: There is a lot of evidence that we may not be able to trust the media created by AI because it can be used to create fake or misleading information. This could make us suspicious of anything we see on social media.

Conclusion

AI does offer the possibility of new creative tools that will be available to all. While this will bring benefits to creative people, social media may be flooded with all sorts of AI-generated content that may mislead users or make them wary of using social media.

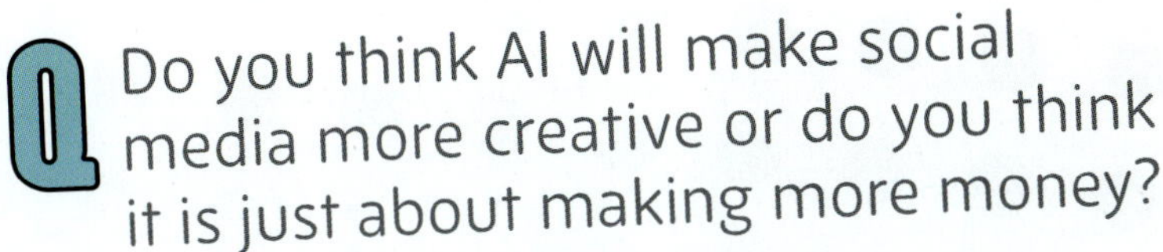

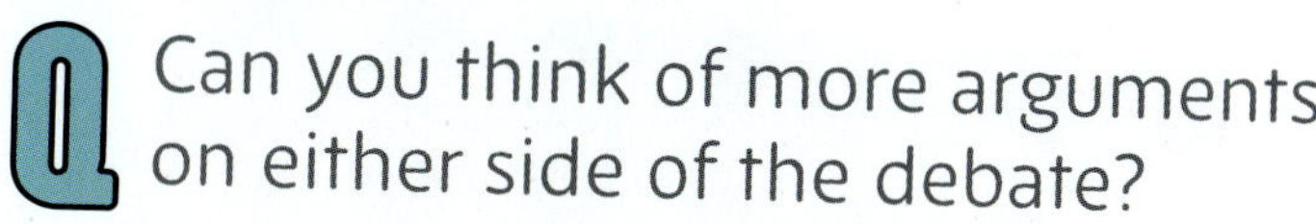

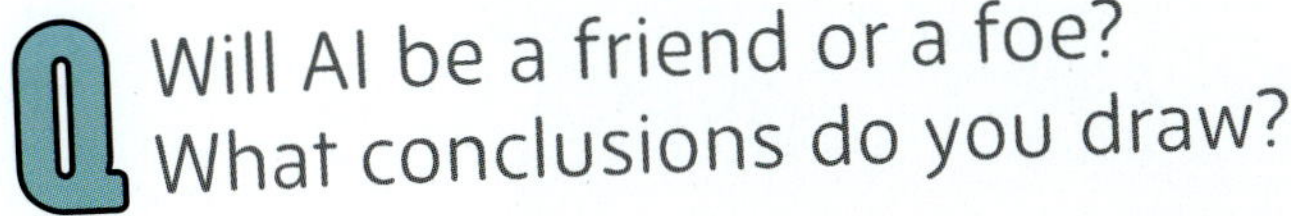

CHAPTER 6

Friend or Foe?

AI has already played a part in changing the way we find entertainment through recommendation algorithms and video games, and in developing video games. AI has also played a huge part in the development of social media, which plays such a big role in many people's lives today. AI technology is developing so quickly that the changes it will bring to entertainment are difficult to predict. There will be winners and losers from the AI revolution.

WHO WILL BE THE WINNERS?

How could AI improve your favorite entertainment? Maybe you're looking forward to playing games with realistic characters controlled by AI. Maybe you think you could use AI to create backing tracks for your own music, or to make videos to share on social media. AI will bring new experiences to all these areas of entertainment.

Producers of entertainment such as movies and video games will be able to use AI to speed up production and to cut costs as AI will take over much of the technical work needed on these projects.

Social media companies have been leading the development and use of AI because these huge corporations stand to make some of the biggest gains. Social media is all about gathering and using data. AI will provide a lot of opportunities to do this and for social media companies to constantly learn more and more about their users.

WHO WILL BE THE LOSERS?

The AI revolution will not benefit everyone equally and there will certainly be losers. Many creative people including writers, actors, and musicians are worried that AI could take away much of their work. For example, AI could create the background music for TV and movies, taking over the work of an orchestra or group of musicians.

Creative people are particularly angry that their work is often being used to train AI tools. AI requires millions of pieces of data so it can learn how to produce writing and images that match up to those created by humans. To do this, it is using writing, artwork, and music created by humans but without them receiving any payment.

This has led to calls for more control of how AI uses data. AI can bring benefits to all of us, but there are also risks. If we rely too much on AI, we may end up with entertainment that feels like it was made by a robot. There is also the risk that AI could get too powerful and difficult to control. Another risk is that AI will provide too much entertainment through gaming platforms and social media. We may find it difficult to switch off and spend time with other people.

Great entertainment can be shared with friends and family. If AI can improve our media choices, this part of our lives could be boosted.

The Debate:

AI Will Change All Entertainment

We know that AI will change the way entertainment is produced and will enable us to generate new content more easily. Perhaps AI will completely change every form of entertainment we enjoy. There are arguments both for and against this scenario. Let's take a look at some of them.

AGREE

Personal entertainment: AI will use data from social media and other sources to understand more about each of us than ever before. This will mean that our entertainment can be more personal to each individual, as social media feeds are now more personal.

Growth of new media: Social media has developed since 2000. AI will help drive further developments of more immersive online experiences, which are being called the metaverse. This new world of immersive experience has great potential and could provide people with a wealth of fun and enriching new forms of entertainment.

Less human involvement: AI will have a big part to play in the production of all entertainment, including movies, music, and games. Humans will still have a part to play but the role of people will be more focused on coming up with new, more creative ideas.

DISAGREE

Human creativity: AI may be able to replicate human ideas but it is less successful at creating new ones. Good entertainment is based on human creativity and that is unlikely to change any time soon.

New and old: AI may provide new ways of creating entertainment and we may watch it in a different way, but we will continue to be excited by great stories, visual spectacles, and music that speaks to us, just as we always have been in the past. That is unlikely to change.

Shared experiences: We may be happy for some entertainment to be personal, but entertainment is also about shared experiences at a concert, a movie, or even a sports event. That's something that AI cannot provide. We need other people to share our experiences.

Conclusion

AI will change the way entertainment is produced and the way we access entertainment. Some areas, such as gaming, may be changed more than others. However, many of the things that we look for as entertainment rely on human involvement and shared experiences. Hopefully those things will continue for a long time to come.

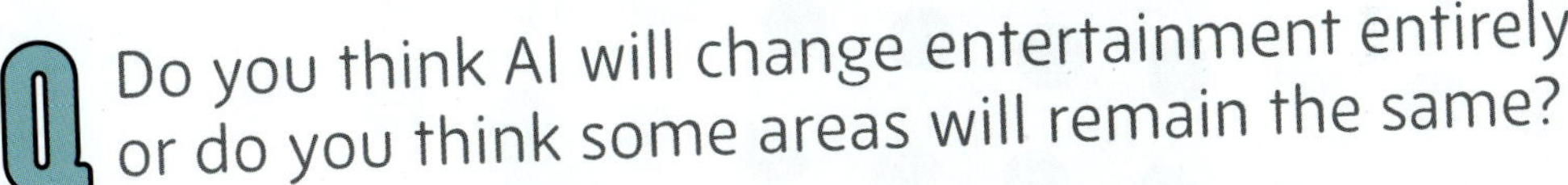

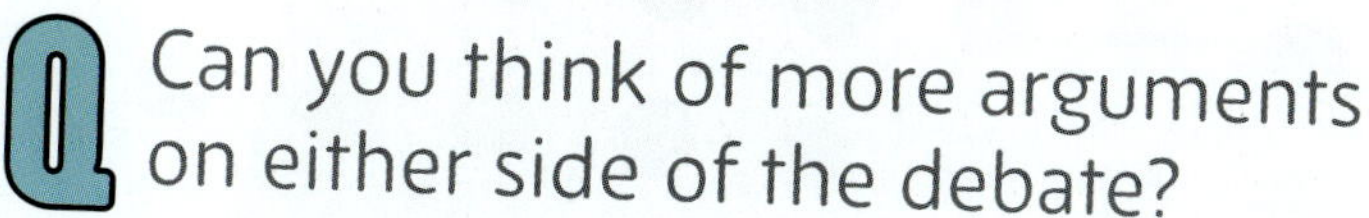

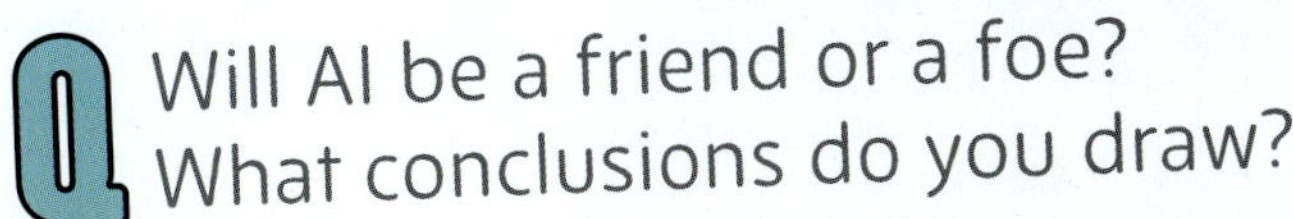

The Ethics of AI

AI could have far-reaching effects on the world of entertainment. This also raises ethical questions, outlined below, for corporations and governments, and for all of us who will use AI in the future.

- Human versus machine: How happy are we for computers to make our entertainment? Would a movie written and acted by AI be a good idea? Where do we draw the line with AI?
- Training of AI: Is it OK for the work of artists, writers, and musicians to be used to provide the data for training AI? This could enable AI to replicate all of their work.
- Privacy: If you use social media, AI is gathering data about you. Should we know more about how this data is being used?

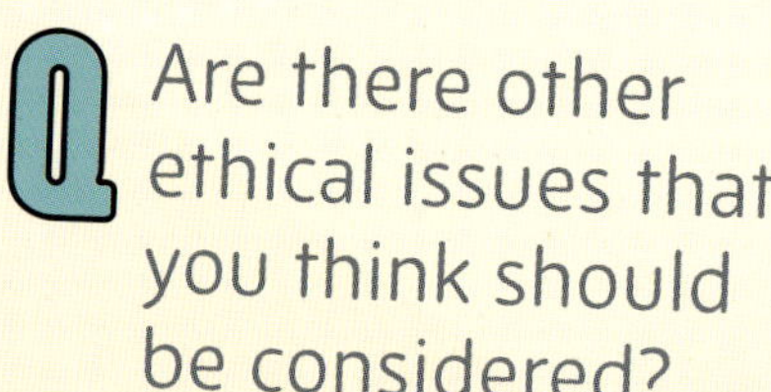

Q Are there other ethical issues that you think should be considered?

- Control and transparency: Will we be told when an image or video has been produced using AI? Shouldn't we know this?
- Fakes and hallucinations: How can we spot fake information or pictures created by AI, and should there be more checks and controls to stop this happening?

Q What rules would you put in place to ensure that AI will have a positive impact on entertainment?

Find Out More

BOOKS

Bathie, Holly. *Social Media Survival Guide* (Usborne Life Skills). Usborne Publishing, 2024.

Enz, Tammy. *Artificial Intelligence and Entertainment: 4D An Augmented Reading Experience* (The World of Artificial Intelligence 4D). Capstone Press, 2019.

Kuromiya, Jun. *The Future of Entertainment* (Searchlight Books—Future Tech). Lerner Publishing Group, 2020.

ONLINE

This video helps explain how AI systems learn:
youtu.be/R9OHn5ZF4Uo

Discover how AI impacts entertainment at:
youtu.be/AjEgejgoO9O?si=7BS-h-6bfPSFJjMx

Find out why your computer learns faster and games better than you might think:
www.mckinsey.com/featured-insights/mckinsey-for-kids/game-on-why-your-computer-learns-faster-and-games-better-than-you-think

To find out more about AI you can also search for websites of companies involved in AI such as OpenAI, the creator of ChatGPT, and Google's DeepMind.

Publisher's note to educators and parents:
All the websites featured above have been carefully reviewed to ensure that they are suitable for students. However, many websites change often, and we cannot guarantee that a site's future contents will continue to meet our high standards of educational value. Please be advised that students should be closely monitored whenever they access the Internet.

Glossary

algorithms processes or sets of rules to be followed for a computer to solve a problem or perform a task

artificial describes something made by humans and not naturally occurring

audiobooks books that can be listened to rather than read

avatars computer-generated images of people, such as used in a game or online communication

beneficial bringing benefit

bias being in favor of one argument or group of people, which may be unfair to others

chatbot computer software designed to communicate with human users online

cloned created an exact copy

computer-generated images (CGI) visual effects that are created by computers

creativity the use of imagination or original ideas to create something

debate argument or discussion about a particular subject, in which arguments are given for and against the main question

deepfakes forms of AI that use deep learning to create fake images

discrimination the unfair treatment of a group of people because of bias

documentary a nonfiction film about real events or people

ethical relating to moral rules that decide how people behave

facial gestures movements of the face that help show emotion

foe an enemy

generative AI software that is able to create new content such as documents or images using AI

hallucinate to see something that is not actually there, now used to describe incorrect information presented as fact by AI

hardware the physical parts of a computer or electronic system

immersive designed to make people, such as game players, feel deeply involved in something

mass entertainment TV shows, movies, and other forms of entertainment that are designed to appeal to a lot of people

metaverse an idea that in the future the Internet becomes one big virtual world

microprocessors parts of a computer in the form of a single chip that can perform calculations and process program commands

movie sets the scenery and props, or objects needed for a scene, that are arranged for shooting a movie

narrate to tell a story or give an account of something by speaking it

original new and unlike anything that has been done before

pandemic the outbreak of an infectious disease that affects much of the world at the same time

permission to agree to something or allow it to happen

predicting saying what will happen in the future

realistic as similar as possible to the real world

replicate to copy or reproduce

robots machines that can replicate human functions or movements automatically

scripts the written texts of plays or movies created by a writer

software programs or instructions that affect how a computer operates

special effects visual effects used in movies and TV, often created by computers

stereotypes oversimplified images of people or a group of people

streaming watching or listening to entertainment direct from the Internet, rather than stored on another device

strike when workers stop work to protest about pay or other workplace issues

Index

About the Author

Nick Hunter is a highly experienced children's book author, who has written countless titles on many subjects, from history and science through social studies and geography. In writing this book he has discovered that AI is an incredibly powerful technology that has the potential to bring great benefits to entertainment if we manage its potential risks.